Mrs. Wiggles and the Numbers

PATTERNS

words and pictures by
LISA KONKOL

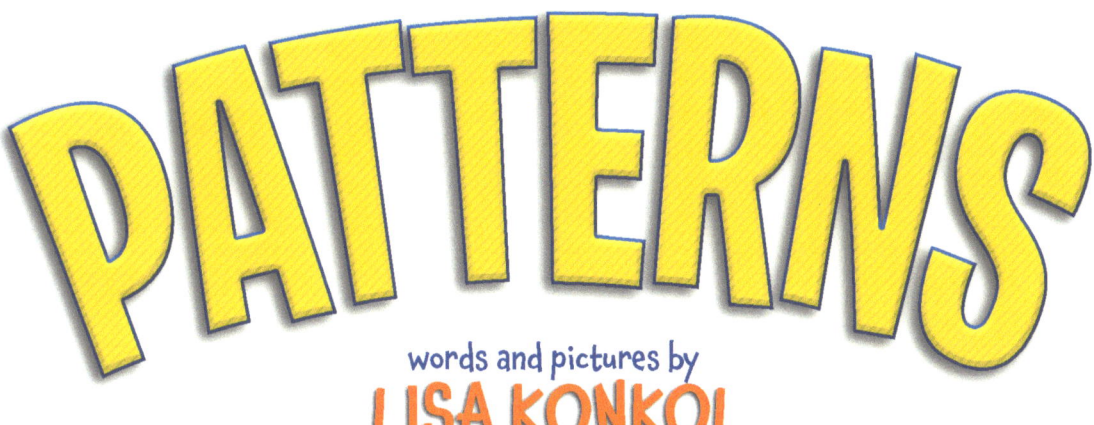

Baa Baa Books, LLC
Madison, Wisconsin

ALSO IN THIS SERIES:
Mrs. Wiggles and the Numbers

Scan the code to visit LisaKonkol.com to download free Mrs. Wiggles activities!

For our grandchildren, with love.

Mrs. Wiggles and the Numbers: Patterns

Text and illustrations copyright © 2025 by Lisa Konkol
No part of this work may be reproduced or transmitted in any form or by any means, electronic or mechanical, including photocopying and recording, or by any information storage or retrieval system, except as may be expressly permitted by the 1976 Copyright Act or in writing from the publisher. Printed in China.
For permission requests contact MrsWiggles@lisakonkol.com.

Books published by Baa Baa Books may be purchased in bulk at special discounts for sales promotion, corporate gifts, fund-raising, or educational purposes. Special editions can also be created to specifications.
For details, contact Special Sales Dept., info@baabaabooks.shop.

Cover and interior design by Lisa Konkol and Tamara Dever, TLCBookDesign.com.

Visit our website at LisaKonkol.com

Softcover: ISBN 978-1-7359196-5-2
Hardcover: ISBN 978-1-7359196-6-9

A special thank you to my critique group and my husband Steve.

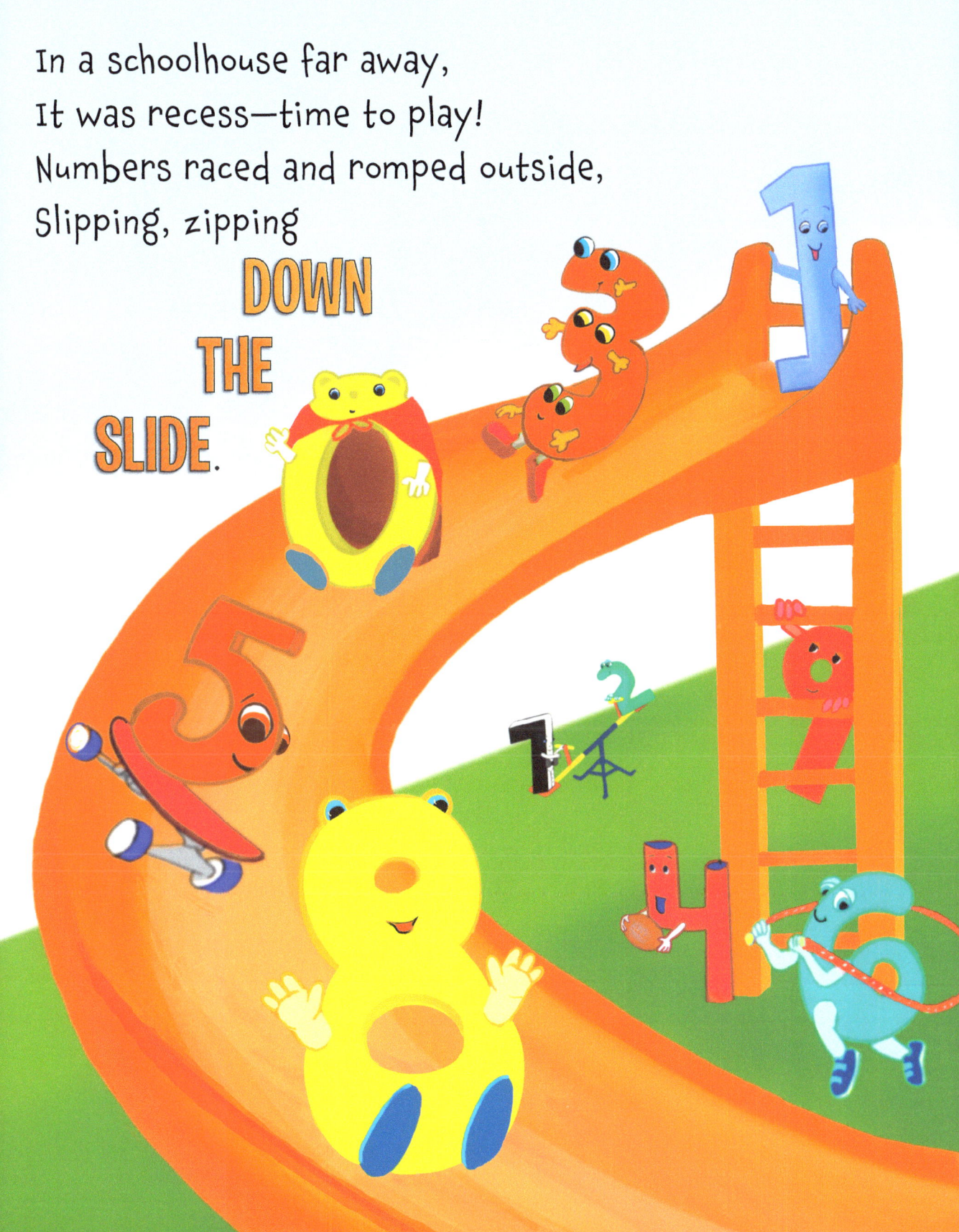

In a schoolhouse far away,
It was recess—time to play!
Numbers raced and romped outside,
Slipping, zipping DOWN THE SLIDE.

With a wink and giant **SWOOP**, Teacher hollered, "Line up, group!"

Patterns can be heard and seen.
Let me show you what I mean."

Zero sounded somewhat scared. "I am not at all prepared."

Seven gave a calm reply. "You and I can always try."

"See the garden straight ahead..."

"There's a pattern!"
Seven said.

In the pasture, white and brown,
Standing tall and sitting down.
"Patterns!" shouted Number One.
"Finding them is lots of fun!"

In the pond, a boisterous C-R-O-A-A-A-K.
Even Number Nine awoke!

Frogs with stripes and some with spots.
Lots of frogs with polka dots.

Someone hollered overhead.
"I'm up here!" a voice then said.
Mrs. Wiggles looked up high.
Who was waving from the sky?

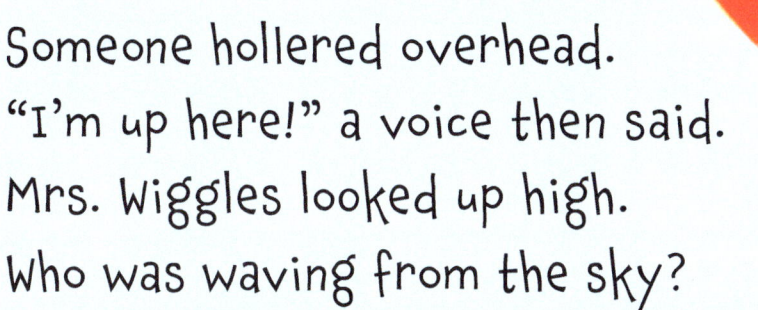

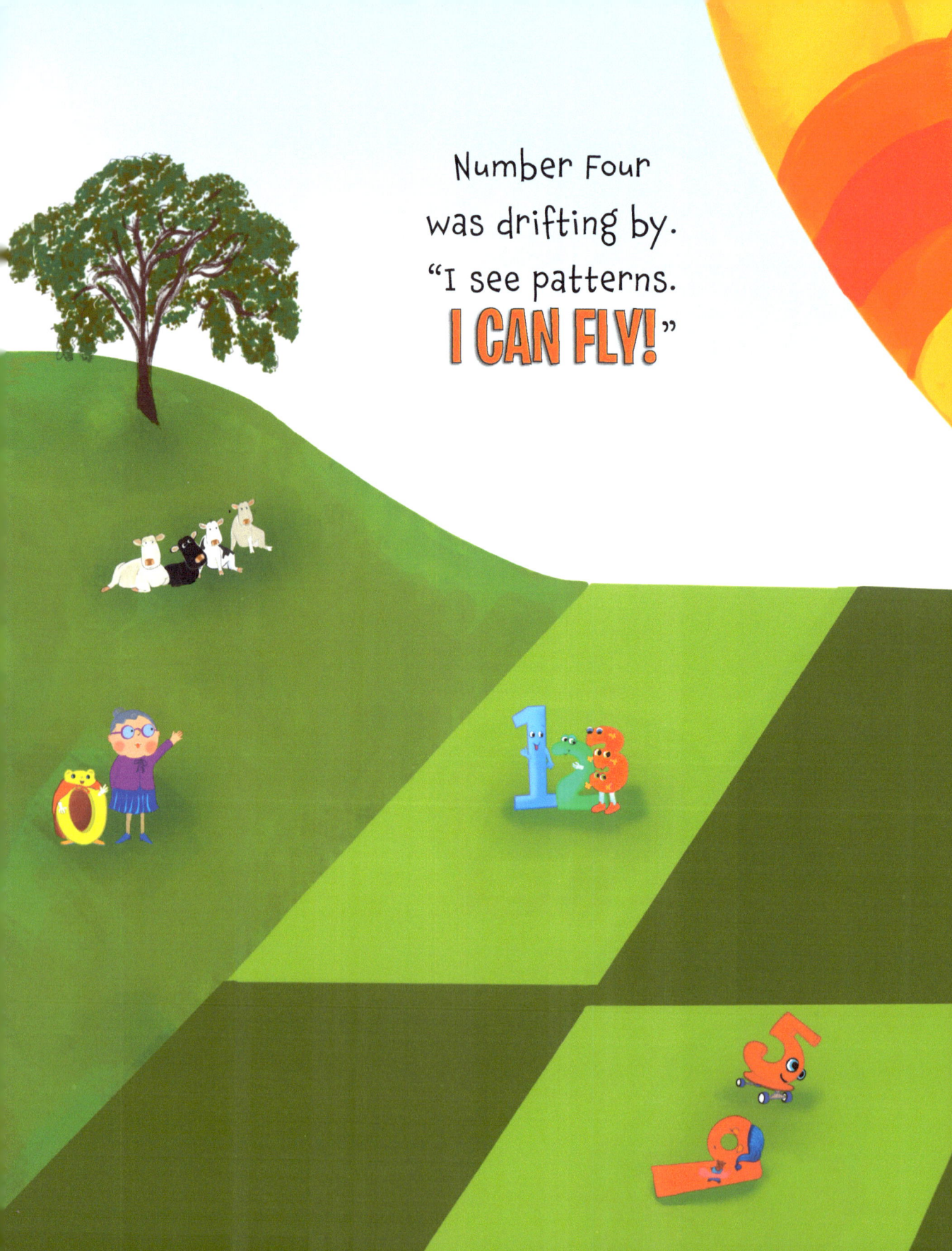

Number Four was drifting by. "I see patterns. **I CAN FLY!**"

"Hmmm," Mrs. Wiggles said.
"What's inside this barnyard shed?"

Lots of chickens all around.
In the coop and on the ground.
Hens were dark, and hens were light...

Speckled eggs were nestled tight.

"Made a pattern!" Zero said.
"Thought a lot and used my head."

"WE are patterns on the wall.
Numbers are the best of all!"

"All together, count by tens.
Numbers make the best of friends."

"**WOW!** One hundred. Huge amount!
Number patterns help us count."

Mrs. Wiggles rang her bell.
"Numbers, you did very well."

www.ingramcontent.com/pod-product-compliance
Lightning Source LLC
Chambersburg PA
CBHW041525070526
44585CB00002B/84